Chapter 8

Two-Digit Addition and Subtraction

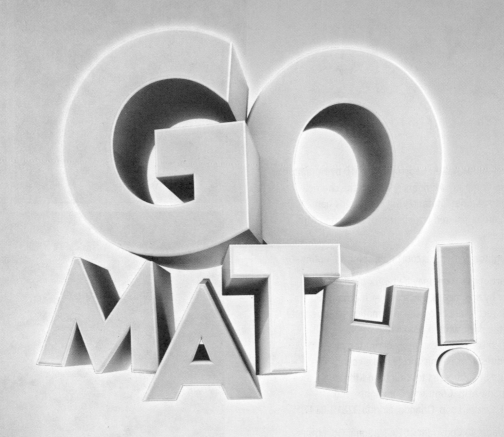

Made in the United States
Text printed on 100%
recycled paper

Houghton
Mifflin
Harcourt

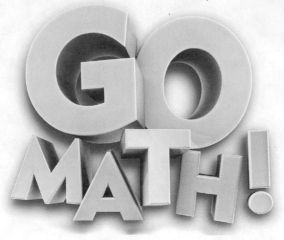

GO MATH!

Printed in the U.S.A.

ISBN 978-0-544-34192-0

16 0928 19

4500746728   C D E F G

Dear Students and Families,

Welcome to **Go Math!**, Grade 1! In this exciting mathematics program, there are hands-on activities to do and real-world problems to solve. Best of all, you will write your ideas and answers right in your book. In **Go Math!**, writing and drawing on the pages helps you think deeply about what you are learning, and you will really understand math!

By the way, all of the pages in your **Go Math!** book are made using recycled paper. We wanted you to know that you can Go Green with **Go Math!**

Sincerely,

The Authors

Made in the United States
Text printed on 100% recycled paper

GO MATH!

Authors

Juli K. Dixon, Ph.D.
Professor, Mathematics Education
University of Central Florida
Orlando, Florida

Edward B. Burger, Ph.D.
President, Southwestern University
Georgetown, Texas

Steven J. Leinwand
Principal Research Analyst
American Institutes for
  Research (AIR)
Washington, D.C.

Contributor

Rena Petrello
Professor, Mathematics
Moorpark College
Moorpark, California

Matthew R. Larson, Ph.D.
K-12 Curriculum Specialist for
  Mathematics
Lincoln Public Schools
Lincoln, Nebraska

Martha E. Sandoval-Martinez
Math Instructor
El Camino College
Torrance, California

English Language Learners Consultant

Elizabeth Jiménez
CEO, GEMAS Consulting
Professional Expert on English
  Learner Education
Bilingual Education and
  Dual Language
Pomona, California

Number and Operations in Base Ten

 Critical Area Developing understanding of whole number relationships and place value, including grouping in tens and ones

 8 **Two-Digit Addition and Subtraction** 433

COMMON CORE STATE STANDARDS

1.OA Operations and Algebraic Thinking
Cluster C Add and subtract within 20.
1.OA.C.6

1.NBT Number and Operations in Base Ten
Cluster C Use place value understanding and properties of operations to add and subtract.
1.NBT.C.4
1.NBT.C.6

GO DIGITAL

Go online! Your math lessons are interactive. Use *iTools*, Animated Math Models, the Multimedia *eGlossary*, and more.

Chapter 8 Overview

In this chapter, you will explore and discover answers to the following **Essential Questions**:

- How can you add and subtract two-digit numbers?
- What ways can you use tens and ones to add and subtract two-digit numbers?
- How can making a ten help you add a two-digit number and a one-digit number?

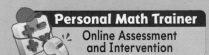 **Personal Math Trainer**
Online Assessment and Intervention

CRITICAL AREA REVIEW PROJECT NUMBERS AROUND US: *www.thinkcentral.com*

FOR MORE PRACTICE
GO TO THE
Personal Math Trainer

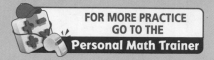

Practice and Homework

Lesson Check and Spiral Review in every lesson

Two-Digit Addition and Subtraction

Curious About Math with

Curious George

There are 4 boxes of oranges on a table. Each box holds 10 oranges. How many oranges are there?

✓ Show What You Know

Personal Math Trainer
Online Assessment
and Intervention

Add and Subtract

Use ■ and ■ to add. Write the sum.
Break apart ■ to subtract.
Write the difference. (K.OA.A.1)

1.

$4 + 1 =$ _____

$5 - 1 =$ _____

Count Groups to 20

Circle groups of 10. Write how many. (1.NBT.A.1)

2. ⭐⭐⭐⭐⭐⭐⭐
⭐⭐⭐⭐⭐⭐⭐

_____

3. ⭐ ⭐
⭐ ⭐
⭐ ⭐
⭐ ⭐ ⭐
⭐ ⭐ ⭐

_____

Use a Hundred Chart to Count

Touch and count. Shade the last
number counted. (1.NBT.A.1)

4. Start at 1 and count to 20.

5. Start at 30 and count to 56.

6. Start at 77 and count to 93.

1	2	3	4	5	6	7	8	9	10
11	12	13	14	15	16	17	18	19	20
21	22	23	24	25	26	27	28	29	30
31	32	33	34	35	36	37	38	39	40
41	42	43	44	45	46	47	48	49	50
51	52	53	54	55	56	57	58	59	60
61	62	63	64	65	66	67	68	69	70
71	72	73	74	75	76	77	78	79	80
81	82	83	84	85	86	87	88	89	90
91	92	93	94	95	96	97	98	99	100

This page checks understanding of important skills needed
for success in Chapter 8.

Vocabulary Builder

Visualize It

Sort the review words from the box.

Put Together

_____

_____

_____

Take Apart

_____

_____

_____

Understand Vocabulary

Use a review word to complete each sentence.

1. 8 is the _____ for 17 − 9.

2. 17 is the _____ for 8 + 9.

3. When you _____ 4 to 8,
you find the sum.

4. When you _____ 4 from 8,
you find the difference.

Game · Neighborhood Sums

Materials

 · ·

Play with a partner.

1. Put your 🔵 on START.
2. Spin the ⊘. Move that number of spaces.
3. Make a ten to help you find the sum.
4. The other player uses 🟦🟦🟦 to check.
5. If you are not correct, you lose a turn.
6. The first player to get to END wins.

2 4 +8	Move ahead one space.	4 9 +6	4 4 +6	9 1 +6	END
4 6 +3	5 3 +7	9 7 +1	Move back one space.	3 7 +7	5 8 +5
					6 6 +4
START	2 4 +8	Move ahead one space.	6 1 +9	8 8 +2	

Chapter 8 Vocabulary

add

sumar

1

addition sentence

enunciado de suma

3

difference

diferencia

13

fewer

menos

19

more

más

38

subtract

restar

52

subtraction sentence

enunciado de resta

53

sum

suma o total

54

4 + 2 = 6

is an **addition sentence.**

3 + 2 = 5

3 **fewer** 🐦

9 – 4 = 5

The **difference** is 5.

5 – 2 = 3

5 – 3 = 2

There are **more** ⭐.

2 plus 1 is equal to 3.
The **sum** is 3.

9 – 5 = 4

is a **subtraction sentence.**

Bingo

Materials
- 1 set of word cards
- 18 ●

How to Play
Play with a partner.
1. Mix the cards. Put the cards in a pile with the blank side up.
2. Take a card. Read the word.
3. Find the matching word on your bingo board. Cover the word with a ●. Put the card at the bottom of the pile.
4. The other player takes a turn.
5. The first player to cover 3 spaces in a line wins. The line may go across or down.

Word Box

Word Box
add
addition sentence
difference
fewer
more
subtract
subtraction sentence
sum

Player 1

fewer	add	more
addition sentence	BINGO	sum
subtraction sentence	difference	subtract

Player 2

subtraction sentence	addition sentence	fewer
add	BINGO	difference
subtract	more	sum

The Write Way

Reflect

Choose one idea. Draw and write about it.

- Write sentences that include at least two of these terms.

  add   subtract   addition sentence   subtraction sentence

- Explain how you would solve this problem.

  $$53 + 20 = \underline{\hspace{2cm}}$$

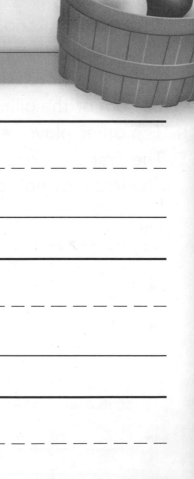

Name _____

Add and Subtract Within 20

Essential Question What strategies can you use to add and subtract?

Common Core — Operations and Algebraic Thinking—1.OA.C.6
MATHEMATICAL PRACTICES
MP1, MP3, MP6

Listen and Draw *Real World*

What is $5 + 4$?
Use a strategy to solve the addition
fact. Draw to show your work.

$5 + 4 =$ _____

Math Talk — MATHEMATICAL PRACTICES 3

FOR THE TEACHER • Have children choose and model a strategy to solve the addition fact. Then have them draw to show their work.

Apply What strategy did you use to find the answer?

Chapter 8

Model and Draw

Think of a strategy you can use
to add or subtract.

What is 14 − 6?

I can use a
related fact.

$\underline{6}$ ⊕ $\underline{8}$ = 14

So, 14 − 6 = $\underline{8}$.

Share and Show

Add or subtract.

1. 5 + 3 = ___   2. 10 − 5 = ___   3. 3 + 6 = ___

4. 12 − 5 = ___   5. 15 − 9 = ___   6. 5 + 7 = ___

7. 8 + 7 = ___   8. 9 − 7 = ___   9. 5 + 5 = ___

10. 12 − 7 = ___   11. 18 − 9 = ___   12. 9 + 4 = ___

13. 2 + 7 = ___   14. 5 − 1 = ___   15. 9 + 1 = ___

16. 7 − 6 = ___   ⊘17. 13 − 4 = ___   ⊘18. 2 + 6 = ___

Name _____

On Your Own

Apply Add or subtract.

19. 14   20.   2   21.   3   22. 14   23.   8   24.   6
    – 5   +10   +3   – 8   +9   –3
    _____   _____   _____   _____   _____   _____

25.   6   26.   2   27.   0   28. 10   29.   9   30.   5
    –5   +8   +5   – 2   +9   –4
    _____   _____   _____   _____   _____   _____

31.   8   32. 10   33.   4   34.   9   35.   1   36. 17
    –8   + 1   +7   –3   +8   – 9
    _____   _____   _____   _____   _____   _____

37. 13   38.   6   39. 10   40. 14   41. 10   42. 11
    – 7   +5   + 2   – 9   +10   – 3
    _____   _____   _____   _____   _____   _____

43. **THINK SMARTER** Jamal thinks of an addition fact. The sum is 15. One addend is 8. What is a fact Jamal could be thinking of?

____ ◯ ____ ◯ ____

Problem Solving • Applications WRITE Math

Solve. Write or draw to explain.

44. THINK SMARTER There are 9 ants on a rock. Some more ants get on the rock. Now there are 18 ants on the rock. How many more ants got on the rock?

_____ more ants

45. GO DEEPER Fill in the blanks. Write a number sentence to solve.

Lin sees _____ bees. Some bees flew away. Now there are _____ bees. How many bees flew away?

___ ◯ ___ ◯ ___

_____ bees

46. THINK SMARTER Write each addition or subtraction problem in the box below the sum or difference.

$7 + 9$ $6 + 1$ $17 - 8$ $14 - 7$ $8 + 8$

7	9	16

 TAKE HOME ACTIVITY • Have your child tell a strategy he or she would use to solve $4 + 8$.

© Houghton Mifflin Harcourt Publishing Company • Image Credits: (b) ©Shutterstock; (t) ©Domiciano Pablo Romero Franco/Alamy

Add and Subtract Within 20

COMMON CORE STANDARD—1.OA.C.6
Add and subtract within 20.

Add or subtract.

1. 6 +0	2. 11 − 2	3. 4 +5	4. 9 +8	5. 4 +10	6. 14 − 9

7. 7 +4	8. 8 −5	9. 10 −10	10. 6 +7	11. 18 − 9	12. 15 − 6

Problem Solving

Solve. Draw or write to explain.

13. Jesse has 4 shells. He finds some more. Now he has 12 shells. How many more shells did Jesse find?

_____ more shells

14. Write an addition or subtraction fact. Then write a strategy you could use to add or subtract.

_____

_____

_____

_____

1. What is the sum?
Write the number.

$$8 + 5 = \underline{}$$

2. What is the difference?
Write the number.

$$11 - 4 = \underline{}$$

Spiral Review (1.NBT.B.3)

3.

Circle the greater number.	Did tens or ones help you decide?	Write the numbers.
43 46	tens ones	_____ is greater than _____. _____ > _____

4.

Circle the number that is less.	Did tens or ones help you decide?	Write the numbers.
69 84	tens ones	_____ is less than _____. _____ < _____

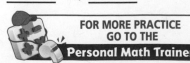

FOR MORE PRACTICE
GO TO THE
Personal Math Trainer

Name _____

Add Tens

Essential Question How can you add tens?

Common Core
Number and Operations in Base Ten—1.NBT.C.4
MATHEMATICAL PRACTICES
MP2, MP7

Listen and Draw (Real World)

Choose a way to show the problem.
Draw a quick picture to show your work.

Math Talk MATHEMATICAL PRACTICES 2

Reasoning Why will there be no ones in your answer when you add 20 + 30?

FOR THE TEACHER • Read the following problems. Barb has 20 baseball cards. Ed has 30 baseball cards. How many baseball cards do they have? Kyle has 40 baseball cards. Kim has 50 baseball cards. How many baseball cards do they have?

Chapter 8

How can you find 30 + 40?

$$30 \quad + \quad 40 \quad = \quad \underline{70}$$

||| ||| ||| |||

_____ tens

Share and Show

Use . Draw to show tens.
Write the sum. Write how many tens.

1. 20 + 40 = _____

_____ tens

2. 30 + 30 = _____

_____ tens

✓3. 40 + 50 = _____

_____ tens

✓4. 50 + 30 = _____

_____ tens

Name _____

On Your Own

MATHEMATICAL PRACTICE ② **Represent a Problem** Draw to show tens.
Write the sum. Write how many tens.

5. 40 + 40 = ____

____ tens

6. 70 + 20 = ____

____ tens

7. 10 + 80 = ____

____ tens

8. 60 + 30 = ____

____ tens

9. **GO DEEPER** Draw two groups of tens you can add to
get a sum of 50. Write the number sentence.

____ ◯ ____ ◯ ____

Problem Solving • Applications  WRITE ▸ Math

10. **THINKSMARTER** Complete the web. Write the missing addend to get a sum of 90.

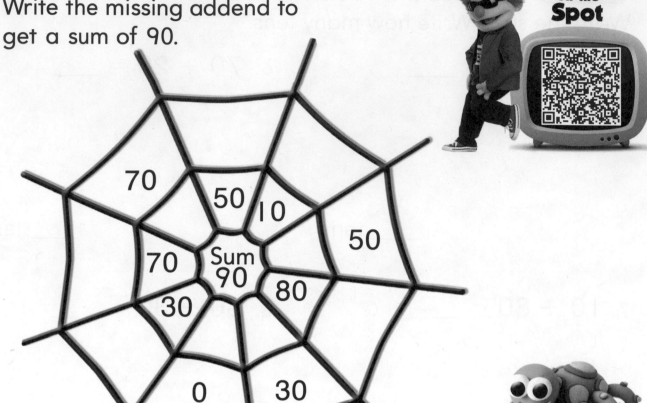

70

50   10

70   Sum
90   50

30   80

0   30

Math on the Spot

11. **THINKSMARTER** Choose all the ways that name the model.

- ○ 4 ones and 3 tens
- ○ 4 tens and 3 tens
- ○ 7 tens
- ○ 70

 TAKE HOME ACTIVITY • Ask your child to explain how to use tens to find 20 + 70.

Name _____

Add Tens

COMMON CORE STANDARD—1.NBT.C.4
Use place value understanding and properties of operations to add and subtract.

Draw to show tens. Write the sum. Write how many tens.

1. 10 + 30 = ____

____ tens

2. 30 + 30 = ____

____ tens

3. 60 + 10 = ____

____ tens

4. 10 + 70 = ____

____ tens

Problem Solving

Draw tens to solve.

5. Drew makes 20 posters. Tia makes 30 posters. How many posters do they make?

____ posters

6. Regina read 40 pages. Alice read 50 pages. How many pages did they read?

____ pages

7. WRITE ▶ Math Choose an addition problem from the spider web on page 446. Draw a quick picture and write the number sentence.

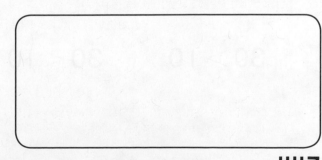

I. What is the sum?
Write the number.

$$20 + 30 = \underline{\quad}$$

2. What is the sum?
Write the number.

$$30 + 10 = \underline{\quad}$$

Spiral Review (1.OA.C.6, 1.NBT.B.3)

3. Write a doubles fact that can help you solve $6 + 5 = 11$.

$$\underline{\quad} + \underline{\quad} = \underline{\quad}$$

4. Circle the number sentences that are true.
Cross out the number sentences that are false.

$$30 > 10 \qquad 30 < 10 \qquad 10 > 30 \qquad 10 < 30$$

**FOR MORE PRACTICE
GO TO THE
Personal Math Trainer**

Name _____

Subtract Tens

Essential Question How can you subtract tens?

Common Core
Number and Operations in
Base Ten—1.NBT.C.6
MATHEMATICAL PRACTICES
MP3, MP4, MP6, MP8

Listen and Draw (Real World)

Choose a way to show the problem.
Draw a quick picture to show your work.

FOR THE TEACHER • Read the following problems.
Tara has 30 seashells. 20 shells are big. The rest
are small. How many small shells does she have?
Sammy has 50 shells. He gives 30 shells to his
friend. How many shells does Sammy have now?

Math Talk MATHEMATICAL PRACTICES 4

Represent How does
your picture show the
first problem?

Chapter 8

four hundred forty-nine **449**

How can you find 80 − 30?

$$80 - 30 = \underline{50}$$

____ tens

Share and Show MATH BOARD

Use ▭▭▭▭▭. Draw to show tens.
Write the difference. Write how many tens.

1. 60 − 20 = ____

____ tens

2. 70 − 30 = ____

____ tens

✓3. 80 − 20 = ____

____ tens

✓4. 90 − 40 = ____

____ tens

Name _____

MATHEMATICAL PRACTICE 6 **Make Connections** Draw to show tens.
Write the difference. Write how many tens.

5. 80 – 40 = ____

_____ tens

6. 90 – 70 = ____

_____ tens

7. 70 – 50 = ____

_____ tens

8. 30 – 30 = ____

_____ tens

THINK SMARTER Solve.

9. Jeff has 40 pennies. He gives some to Jill. He has 10 pennies left. How many pennies does Jeff give to Jill?

Math on the Spot

_____ pennies

TAKE HOME ACTIVITY • Ask your child to explain how to use tens to find 90 – 70.

Name _____

Concepts and Skills

Add or subtract. (1.OA.C.6)

1. 4
  +8

2. 15
  − 7

3. 9
  −6

4. 3
  +1

5. 10
  + 6

6. 11
  − 2

Use ▭ ▪. Draw to show tens.
Write the sum. Write how many tens. (1.NBT.C.4)

7. $30 + 50 =$ ___

8. $40 + 20 =$ ___

____ tens

____ tens

Use ▭ ▪. Draw to show tens.
Write the difference. Write how many tens. (1.NBT.C.6)

9. $90 − 20 =$ ___

10. $60 − 40 =$ ___

____ tens

____ tens

11. **THINK SMARTER** Mike has 60 marbles.
He gives 20 to Kathy. How many
marbles does Mike have left?
Show your work. (1.NBT.C.6)

_____ marbles

Name _____

Subtract Tens

Draw to show tens. Write the difference. Write how many tens.

COMMON CORE STANDARD—1.NBT.C.6
Use place value understanding and properties of operations to add and subtract.

1. 40 − 10 = ____

____ tens

2. 80 − 40 = ____

____ tens

Problem Solving Real World

Draw tens to solve.

3. Mario has 70 baseball cards.
   He gives 30 to Lisa.
   How many baseball cards
   does Mario have left?

____ baseball cards

4. **WRITE Math** Draw a picture to
   show how to solve 50 − 40.

1. What is the difference?
Write the number.

$$60 - 20 = \underline{\hspace{2em}}$$

2. What is the difference?
Write the number.

$$70 - 30 = \underline{\hspace{2em}}$$

Spiral Review (1.OA.C.6, 1.NBT.B.3)

3. Use ○ ● and a ten frame. Show
both addends. Draw to make ten.
Then write a new fact. Add.

$$\begin{array}{r} 9 \\ + 4 \\ \hline \end{array}$$

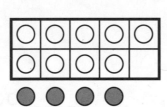

$+$
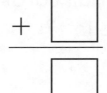

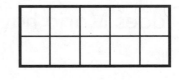

4. Bo crosses out the number cards that are
less than 33 or greater than 38.
What number cards are left?

| 30 | 32 | 36 | 37 | 39 |

Number cards _____ and _____ are left.

© Houghton Mifflin Harcourt Publishing Company

FOR MORE PRACTICE
GO TO THE
Personal Math Trainer

Name _____

Use a Hundred Chart to Add

Essential Question How can you use a hundred chart to count on by ones or tens?

Common Core
Number and Operations in Base Ten—1.NBT.C.4
MATHEMATICAL PRACTICES
MP4, MP5, MP6

Listen and Draw · Real World

Use the hundred chart to solve the problems.

1	2	3	4	5	6	7	8	9	10
11	12	13	14	15	16	17	18	19	20
21	22	23	24	25	26	27	28	29	30
31	32	33	34	35	36	37	38	39	40
41	42	43	44	45	46	47	48	49	50
51	52	53	54	55	56	57	58	59	60
61	62	63	64	65	66	67	68	69	70
71	72	73	74	75	76	77	78	79	80
81	82	83	84	85	86	87	88	89	90
91	92	93	94	95	96	97	98	99	100

Math Talk
MATHEMATICAL PRACTICES 6

Explain how you can use a hundred chart to find each sum.

FOR THE TEACHER • Read the following problems. Alice picks 12 flowers. Then she picks 4 more flowers. How many flowers does Alice pick? Ella picks 10 strawberries. Then she picks 20 more strawberries. How many strawberries does Ella pick?

Model and Draw

Count on a hundred chart
to find a sum.

Start at **24**.
Count on four ones.
25, 26, 27, 28

1	2	3	4	5	6	7	8	9	10
11	12	13	14	15	16	17	18	19	20
21	22	23	24	25	26	27	(28)	29	30
31	32	33	34	35	36	37	38	39	40
41	42	43	44	45	46	47	48	49	50
51	52	53	54	55	56	57	58	59	60
61	62	63	64	65	66	67	68	69	70
(71)	72	73	74	75	76	77	78	79	80
81	82	83	84	85	86	87	88	89	90
91	92	93	94	95	96	97	98	99	100

$$24 + 4 = \underline{28}$$

Start at **31**.
Count on four tens.
41, 51, 61, 71

$$31 + 40 = \underline{71}$$

Share and Show MATH BOARD

Use the hundred chart to add.
Count on by ones or tens.

1. $42 + 7 = \underline{}$

2. $57 + 30 = \underline{}$

☑ 3. $91 + 5 = \underline{}$

☑ 4. $18 + 50 = \underline{}$

On Your Own

1	2	3	4	5	6	7	8	9	10
11	12	13	14	15	16	17	18	19	20
21	22	23	24	25	26	27	28	29	30
31	32	33	34	35	36	37	38	39	40
41	42	43	44	45	46	47	48	49	50
51	52	53	54	55	56	57	58	59	60
61	62	63	64	65	66	67	68	69	70
71	72	73	74	75	76	77	78	79	80
81	82	83	84	85	86	87	88	89	90
91	92	93	94	95	96	97	98	99	100

How can you use the hundred chart to find each sum?

$32 + 5 =$ ___

$48 + 30 =$ ___

MATHEMATICAL PRACTICE 5 Use Appropriate Tools

Use the hundred chart to add.
Count on by ones or tens.

5. $13 + 70 =$ ___

6. $22 + 6 =$ ___

7. $71 + 3 =$ ___

8. $49 + 50 =$ ___

9. $53 + 4 =$ ___

10. $25 + 40 =$ ___

11. **GO DEEPER** Solve. Show your work.

$31 + 20 + 40 =$ ___

Problem Solving • Applications WRITE Math

Choose a way to solve. Draw or write to show your work.

12. **THINK SMARTER** Rae put 20 books away. She put 20 more books away, then 11 more. How many books did Rae put away?

_____ books

Personal Math Trainer

13. **THINK SMARTER +** Use the hundred chart to add. Count on by ones or tens.

$62 + 9 =$ _____

1	2	3	4	5	6	7	8	9	10
11	12	13	14	15	16	17	18	19	20
21	22	23	24	25	26	27	28	29	30
31	32	33	34	35	36	37	38	39	40
41	42	43	44	45	46	47	48	49	50
51	52	53	54	55	56	57	58	59	60
61	62	63	64	65	66	67	68	69	70
71	72	73	74	75	76	77	78	79	80
81	82	83	84	85	86	87	88	89	90
91	92	93	94	95	96	97	98	99	100

Explain how you used the chart to find the sum.

 TAKE HOME ACTIVITY • On a piece of paper, write 36 + 40. Ask your child to explain how to use the hundred chart to count on by tens to find the sum.

458 four hundred fifty-eight

Name _____

Use a Hundred Chart to Add

Common Core
COMMON CORE STANDARD—1.NBT.C.4
Use place value understanding and properties of operations to add and subtract.

Use the hundred chart to add.
Count on by ones or tens.

1. 47 + 2 = _____

2. 26 + 50 = _____

3. 22 + 5 = _____

4.   4 + 85 = _____

1	2	3	4	5	6	7	8	9	10
11	12	13	14	15	16	17	18	19	20
21	22	23	24	25	26	27	28	29	30
31	32	33	34	35	36	37	38	39	40
41	42	43	44	45	46	47	48	49	50
51	52	53	54	55	56	57	58	59	60
61	62	63	64	65	66	67	68	69	70
71	72	73	74	75	76	77	78	79	80
81	82	83	84	85	86	87	88	89	90
91	92	93	94	95	96	97	98	99	100

Problem Solving 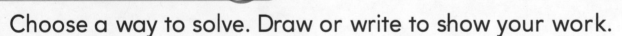 Real World

Choose a way to solve. Draw or write to show your work.

5. 17 children are on the bus.
   Then 20 more children get
   on the bus. How many
   children are on the bus now?   _____ children

6. **WRITE** **Math** Write a number   _____
   sentence to add 6 ones to 21.
   Write a number sentence to   _____
   add 6 tens to 21.
                                        _____

1. What is the sum?
   Write the number.

   $42 + 50 =$ ___

1	2	3	4	5	6	7	8	9	10
11	12	13	14	15	16	17	18	19	20
21	22	23	24	25	26	27	28	29	30
31	32	33	34	35	36	37	38	39	40
41	42	43	44	45	46	47	48	49	50
51	52	53	54	55	56	57	58	59	60
61	62	63	64	65	66	67	68	69	70
71	72	73	74	75	76	77	78	79	80
81	82	83	84	85	86	87	88	89	90
91	92	93	94	95	96	97	98	99	100

2. What is the sum?
   Write the number.

   $11 + 8 =$ ___

Spiral Review (1.OA.D.8, 1.NBT.C.5)

3. Use mental math.
   What number is ten less than 52?
   Write the number.

   ___

4. Write an addition fact that helps
   you solve $16 - 9$.

   ___ + ___ = ___

**FOR MORE PRACTICE
GO TO THE
Personal Math Trainer**

Name _____

Use Models to Add

Essential Question How can models help you add ones or tens to a two-digit number?

Common Core

Number and Operations in Base Ten—1.NBT.C.4
MATHEMATICAL PRACTICES
MP4, MP6

Listen and Draw Real World

Draw to show how you can find the sum.

$14 + 5 = $ _____

Math Talk

MATHEMATICAL PRACTICES 4

Model Explain how you found the sum.

FOR THE TEACHER • Read the following problem. Amir counts 14 cars as they go by. Then he counts 5 more cars. How many cars does Amir count?

Add ones to a two-digit number.

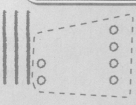

32 + 4 = _36_

Add tens to a two-digit number.

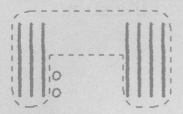

32 + 40 = _72_

Share and Show MATH BOARD

Use . Draw to show how to add the ones. Write the sum.

1. 27 + 2 = ____

☑ 2. 41 + 5 = ____

Use . Draw to show how to add the tens. Write the sum.

3. 13 + 50 = ____

☑ 4. 28 + 30 = ____

On Your Own

 **Use Models**

Use ▭▭▭ ▪ and your MathBoard.
Add the ones or tens. Write the sum.

5. $65 + 3 =$ ___

6. $81 + 8 =$ ___

7. $54 + 20 =$ ___

8. $32 + 10 =$ ___

9. $95 + 2 =$ ___

10. $25 + 60 =$ ___

11. $2 + 54 =$ ___

12. $70 + 29 =$ ___

GO DEEPER Make a sum of 45. Draw a quick picture. Write the number sentence.

13. Add ones to a two-digit number.

___ + ___ = 45

14. Add tens to a two-digit number.

___ + ___ = 45

Problem Solving • Applications

 Real World

WRITE) Math

Choose a way to solve. Draw or write to show your work.

15. Rita picks 63 strawberries. Then she picks 30 more. How many strawberries does Rita pick?

_____ strawberries

16. **THINK SMARTER** Kenny planted two rows of corn. He used 20 seeds in each row. He has 18 seeds left. How many seeds of corn did Kenny have?

_____ seeds

17. There are 7 oak trees and 32 pine trees in the park. How many trees are in the park?

_____ trees

18. **THINK SMARTER** Use the model. Draw to show how to add the tens.

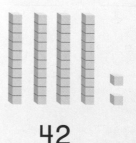  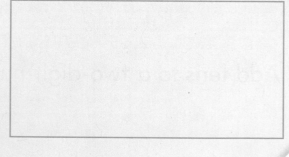

42   +   20   =   _____

 TAKE HOME ACTIVITY • Give your child the addition problems 25 + 3 and 25 + 30. Ask your child to explain how to solve each problem.

Use Models to Add

COMMON CORE STANDARD—1.NBT.C.4
Use place value understanding and properties of operations to add and subtract.

Use ▭▭ ▯ and your MathBoard.
Add the ones or tens. Write the sum.

1. $44 + 5 =$ ___

2. $16 + 70 =$ ___

3. $78 + 20 =$ ___

4. $52 + 7 =$ ___

5. $2 + 13 =$ ___

6. $73 + 4 =$ ___

7. $65 + 3 =$ ___

8. $20 + 25 =$ ___

9. $49 + 30 =$ ___

10. $81 + 8 =$ ___

Problem Solving Real World

Solve. Draw or write to explain.

11. Maria has 21 marbles.
She buys a bag of 20 marbles.
How many marbles does
Maria have now?

___ marbles

12. Write a story problem about 40 apples and 17 pears.

_____

_____

_____

1. What is the sum?
Write the number.

$$62 + 30 = \underline{\hspace{1cm}}$$

2. What is the sum?
Write the number.

$$37 + 2 = \underline{\hspace{1cm}}$$

Spiral Review (1.OA.C.6, 1.NBT.A.1)

3. Write two ways to make 15.

$$\underline{\hspace{1cm}} + \underline{\hspace{1cm}} = 15$$

$$\underline{\hspace{1cm}} + \underline{\hspace{1cm}} = 15$$

4. What number does the model show?

_____

FOR MORE PRACTICE
GO TO THE
Personal Math Trainer

Name _____

Make Ten to Add

Essential Question How can making a ten help you add a two-digit number and a one-digit number?

Common Core **Number and Operations in Base Ten—1.NBT.C.4**
MATHEMATICAL PRACTICES
MP2, MP5

Listen and Draw Real World   Hands On

Use ▭▭▭▭ ▪. Draw to show how you can find the sum.

21 + 6 = _____.

Math Talk   MATHEMATICAL PRACTICES 5

Use Tools Explain how your model shows the sum of 21 + 6.

FOR THE TEACHER • Read the following problem. Sally has 21 stickers in her sticker book. She gets 6 more stickers. How many stickers does Sally have now?

© Houghton Mifflin Harcourt Publishing Company

Make a ten to find $37 + 8$.

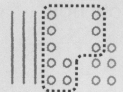

What can I add to 7 to make 10?

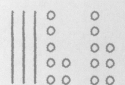

$37 + 8$

$37 + 3 + 5$

$40 + 5$

$\underline{40} + \underline{5} = \underline{45}$

So, $37 + 8 = \underline{45}$.

Share and Show MATH BOARD

Use ▭▭▭▭ . Draw to show how you make a ten. Find the sum.

✓1. $49 + 3 = ?$

_____ + _____ = _____

So, $49 + 3 = $ _____.

Name _____

MATHEMATICAL PRACTICE ⑤ **Use a Concrete Model**

Use ▭▭▭▭▭ ▪. Draw to show how you make a ten. Find the sum.

2. $39 + 7 =$ _____

3. $72 + 9 =$ _____

4. $58 + 5 =$ _____

THINK SMARTER Solve. Write the numbers.

5. $46 + 7$

$46 + \boxed{} + 3$

$\boxed{} + 3$

So, $46 + 7 =$ _____.

6. $53 + 8$

$53 + \boxed{} + 1$

$\boxed{} + 1$

So, $53 + 8 =$ _____.

Math on the Spot

Problem Solving • Applications    Math

Choose a way to solve. Draw
or write to show your work.

7. **THINK SMARTER** Koby puts 24 daisies
and 8 tulips in a vase. How many
flowers are in the vase?

_____ flowers

8. **GO DEEPER** There are 27 ducklings
in the water. 20 of them come
out of the water. How
many ducklings are
still in the water?

_____ ducklings

9. Write the missing addend.

$$46 + \boxed{} = 52$$

10. **THINK SMARTER** Use the model. Draw to
show how to make a ten.

34   +   8   =   _____

TAKE HOME ACTIVITY • Ask your child to explain how to
find the sum for 25 + 9.

Make Ten to Add

Common Core **COMMON CORE STANDARD—1.NBT.C.4**
Use place value understanding and properties of operations to add and subtract.

Use . Draw to show how you make a ten. Find the sum.

1. 26 + 5 = _____

....

2. 68 + 4 = _____

Problem Solving *Real World*

Choose a way to solve. Draw or write to show your work.

3. Debbie has 27 markers. Sal has 9 markers. How many markers do they have?

_____ markers

4. **WRITE** Math Use words or pictures to explain how to solve 44 + 7.

1. What is the sum?
   Write the number.

$$47 + 6 = \underline{}$$

2. What is the sum?
   Write the number.

$$84 + 8 = \underline{}$$

Spiral Review (1.OA.D.7, 1.NBT.A.1)

3. What number does the
   model show?
   Write the number.

4. Write a number to make the sentence true.

$$5 + 4 = 10 - \underline{}$$

FOR MORE PRACTICE
GO TO THE
Personal Math Trainer

Name _____

Use Place Value to Add

Essential Question How can you model tens and ones to help you add two-digit numbers?

Common Core
Number and Operations in Base Ten—1.NBT.C.4
MATHEMATICAL PRACTICES
MP1, MP2, MP6, MP7

Listen and Draw Real World   Hands On

Model the problem with .
Draw a quick picture to show your work.

Tens	Ones

FOR THE TEACHER • Read the following problem. Cameron has 30 old stamps and 25 new stamps. How many stamps does Cameron have?

Math Talk
MATHEMATICAL PRACTICES
Describe How many tens? How many ones? How many in all?

How can you use tens and ones to add?

35
+38

Tens	Ones

3 tens + 5 ones
3 tens + 8 ones
_____
__6__ tens + __13__ ones

__60__ + __13__ = __73__

35
+38
___
73

Share and Show MATH BOARD

Draw a quick picture.
Use tens and ones to add.

 1.

Tens	Ones

81
+14

8 tens + 1 one
1 ten + 4 ones
_____
____ tens + ____ ones

___ + ___ = ___

81
+14

On Your Own

MATHEMATICAL PRACTICE 6 **Make Connections**

Draw a quick picture. Use tens and ones to add.

2.

Tens	Ones

43
+37

4 tens + 3 ones
3 tens + 7 ones
_____
___ tens + ___ ones

___ + ___ = ___

43
+37

3.

Tens	Ones

62
+23

6 tens + 2 ones
2 tens + 3 ones
_____
___ tens + ___ ones

___ + ___ = ___

62
+23

THINK SMARTER Solve.

4. 28 + 17

28 + ___ + 15

___ + 15 = ___

So, 28 + 17 = ___.

5. 59 + 13

59 + ___ + 12

___ + 12 = ___

So, 59 + 13 = ___.

6. **THINK SMARTER** Draw a quick picture to solve. Han has 37 shells. Jonah has 15 shells. How many shells do they have?

Tens	Ones

___

Problem Solving • Applications    WRITE Math

Tens	Ones

7. **THINK SMARTER** Draw a quick picture to solve. Kim has 24 marbles. Al has 47 marbles. How many marbles do they have?

_____ marbles

8. **GO DEEPER** Choose two addends from 11 to 49. Draw them. Add in any order to solve.

Addend   **Addend**

___ + ___ = ___

___ + ___ = ___

9. **THINK SMARTER** Write the addition that the model shows. Solve.

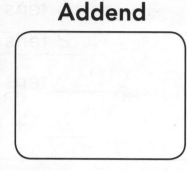

___ + ___ = ___

 TAKE HOME ACTIVITY • Write the numbers 42 and 17. Have your child tell how to find the sum by adding the tens and ones.

Use Place Value to Add

Common Core

COMMON CORE STANDARD—1.NBT.C.4
Use place value understanding and properties of operations to add and subtract.

Draw a quick picture. Use tens and ones to add.

1.

$\begin{array}{r} 31 \\ + 26 \\ \hline \end{array}$

Tens	Ones

3 tens + 1 one
2 tens + 6 ones
_____

____ tens + ____ ones

____ + ____ = ____

$\begin{array}{r} 31 \\ + 26 \\ \hline \end{array}$

2.

$\begin{array}{r} 54 \\ + 34 \\ \hline \end{array}$

Tens	Ones

5 tens + 4 ones
3 tens + 4 ones
_____

____ tens + ____ ones

____ + ____ = ____

$\begin{array}{r} 54 \\ + 34 \\ \hline \end{array}$

Problem Solving

3. Write two addition sentences you can use to find the sum. Then solve.

Addend   **Addend**

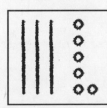   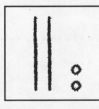

____ + ____ = ____

____ + ____ = ____

4. WRITE Math Write and solve a story problem to add 12 and 18.

_____

_____

1. What is the sum?
   Write the number.

$$\begin{array}{r} 42 \\ +\ 31 \\ \hline \end{array}$$

2. What is the sum?
   Write the number.

$$\begin{array}{r} 23 \\ +\ 12 \\ \hline \end{array}$$

Spiral Review (1.OA.C.6, 1.NBT.B.2)

3. I have 28 cubes. How many tens and ones can I make?

   ____ tens ____ ones

   ____ ten ____ ones

4. What is the sum?
   Write the number.

$$\begin{array}{r} 5 \\ +\ 5 \\ \hline \end{array}$$

© Houghton Mifflin Harcourt Publishing Company

**FOR MORE PRACTICE
GO TO THE
Personal Math Trainer**

Name _____

Problem Solving •
Addition Word Problems

Essential Question How can drawing a picture help you explain how to solve an addition problem?

Common Core — **Number and Operations in Base Ten—1.NBT.C.4**
MATHEMATICAL PRACTICES
MP1, MP2, MP6, MP8

Kelly gets 6 new toy cars.
He already has 18 toy cars.
How many does he have now?

🔑 Unlock the Problem

What do I need to find?

how many _____toy cars_____
Kelly has now

What information do I need to use?

Kelly has ___18___ cars.

He gets ___6___ more cars.

Show how to solve the problem.

_____
- - - - - - - - - - - - - - - - - - -
_____

HOME CONNECTION • Being able to show and explain how to solve a problem helps your child build on their understanding of addition.

Draw and write to solve.
Explain your reasoning.

- What do I need to find?
- What information do I need to use?

1. Aisha picks 60 blueberries to make a pie. Then she picks 12 more to eat. How many blueberries does Aisha pick?

_____ blueberries

_____

_ _ _ _ _ _ _ _ _ _ _ _ _ _ _ _ _ _

_____

2. Yuri collects 21 cans for the school food drive. Leo collects 36 cans. How many cans do Yuri and Leo collect?

_____ cans

_____

_ _ _ _ _ _ _ _ _ _ _ _ _ _ _ _ _ _

_____

Math Talk

MATHEMATICAL PRACTICES 6

Explain the addition strategy you used to solve Exercise I.

Name _____

MATHEMATICAL PRACTICE 2 **Use Reasoning**

Draw and write to solve.

☑3. Tyra sees 48 geese in the field. Then she sees 17 more geese in the sky. How many geese does Tyra see?

_____ geese

_____

_____

☑4. Jade paints 35 circles and 45 triangles in art class. How many shapes does Jade paint?

_____ shapes

_____

_____

5. **THINK SMARTER** It takes 10 hops to get across the yard. How many hops does it take to get across the yard and back?

_____ hops

_____

_____

On Your Own

Choose a way to solve. Draw or write to explain.

6. **THINK SMARTER**   Julian sells 3 books of tickets for the school fair. Each book has 20 tickets. How many tickets does Julian sell?

_____ tickets

7. **GO DEEPER**   I have some red roses and pink roses. I have 14 red roses. I have 8 more pink roses than red roses. How many roses do I have?

_____ roses

Personal Math Trainer

8. **THINK SMARTER +**   Ella sees 27 🧢. She sees 28 🧢. How many 🧢🧢 does Ella see? Circle the number that makes this sentence true.

Ella sees | 48 / 51 / 55 | 🧢🧢 in all.

 TAKE HOME ACTIVITY • Ask your child to solve 16 + 7, 30 + 68, and 53 + 24. Ask him or her to explain how they solved each problem.

Problem Solving • Addition Word Problems

COMMON CORE STANDARD—1.NBT.C.4
Use place value understanding and properties of operations to add and subtract.

Draw and write to solve. Explain your reasoning.

1. Jean has 10 fish. She gets 4 more fish. How many fish does she have now?

   _____

   _____

   _____

   ____ fish

2. Courtney buys 2 bags of apples. Each bag has 20 apples. How many apples does she buy?

   _____

   _____

   _____

   ____ apples

3. John bakes 18 blueberry muffins and 12 banana muffins for the bake sale. How many muffins does he bake?

   _____

   _____

   ____ muffins

4. **WRITE** Math Draw a picture to show how to find 12 + 37.

Lesson Check (1.NBT.C.4)

1. Amy has 9 books about dogs.
She has 13 books about cats.
How many books does she
have about dogs and cats?
Solve. Show your work. Write the number. _____ books

Spiral Review (1.OA.B.3, 1.OA.C.6)

2. What is the sum for $4 + 2 + 4$?
Write the number.

_____

3. Solve. Use the ten frame to make a
ten to help you subtract. Ray has
14 pens. 8 are black. The rest
are blue. How many pens are blue?   _____ blue pens

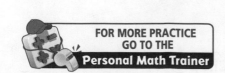

FOR MORE PRACTICE
GO TO THE
Personal Math Trainer

Name _____

Related Addition and Subtraction

Essential Question How can you use a hundred chart to show the relationship between addition and subtraction?

Common Core — Number and Operations in Base Ten—1.NBT.C.4
MATHEMATICAL PRACTICES
MP2, MP3, MP7

Listen and Draw · Real World

Use the hundred chart to solve the problems.

1	2	3	4	5	6	7	8	9	10
11	12	13	14	15	16	17	18	19	20
21	22	23	24	25	26	27	28	29	30
31	32	33	34	35	36	37	38	39	40
41	42	43	44	45	46	47	48	49	50
51	52	53	54	55	56	57	58	59	60
61	62	63	64	65	66	67	68	69	70
71	72	73	74	75	76	77	78	79	80
81	82	83	84	85	86	87	88	89	90
91	92	93	94	95	96	97	98	99	100

Math Talk MATHEMATICAL PRACTICES 3

Apply Describe how you can use a hundred chart to find the sum and the difference.

FOR THE TEACHER • Read the following problems. Trevor collects 38 acorns. He collects 10 more acorns. How many acorns does Trevor have now? Trevor has 48 acorns. He gives 10 acorns to his brother. How many acorns does Trevor have now?

Model and Draw

You can use a hundred chart to find a sum and a difference.

> Start at **29**. Count on four tens.
> **39, 49, 59, 69**

$$29 + 40 = \underline{69}$$

> Start at **69**. Count back four tens.
> **59, 49, 39, 29**

$$69 - 40 = \underline{29}$$

1	2	3	4	5	6	7	8	9	10
11	12	13	14	15	16	17	18	19	20
21	22	23	24	25	26	27	28	29	30
31	32	33	34	35	36	37	38	39	40
41	42	43	44	45	46	47	48	49	50
51	52	53	54	55	56	57	58	59	60
61	62	63	64	65	66	67	68	69	70
71	72	73	74	75	76	77	78	79	80
81	82	83	84	85	86	87	88	89	90
91	92	93	94	95	96	97	98	99	100

Share and Show

Use the hundred chart to add and subtract.
Count on and back by tens.

1. $56 + 20 = \underline{}$

   $76 - 20 = \underline{}$

2. $48 + 50 = \underline{}$

   $98 - 50 = \underline{}$

Name _____

On Your Own

How can you use the hundred chart to find the sum and the difference?

28 + 60 = ___

88 − 60 = ___

1	2	3	4	5	6	7	8	9	10
11	12	13	14	15	16	17	18	19	20
21	22	23	24	25	26	27	28	29	30
31	32	33	34	35	36	37	38	39	40
41	42	43	44	45	46	47	48	49	50
51	52	53	54	55	56	57	58	59	60
61	62	63	64	65	66	67	68	69	70
71	72	73	74	75	76	77	78	79	80
81	82	83	84	85	86	87	88	89	90
91	92	93	94	95	96	97	98	99	100

MATHEMATICAL PRACTICE 7 **Look for a Pattern** Use the hundred chart to add and subtract. Count on and back by tens.

3. 36 + 30 = ___

   66 − 30 = ___

4. 73 + 10 = ___

   83 − 10 = ___

5. 25 + 70 = ___

   95 − 70 = ___

6. 18 + 40 = ___

   58 − 40 = ___

7. **THINK SMARTER** Solve.

There are 73 bees in a hive. 10 bees fly away. Then 10 more bees fly into the hive. How many bees are in the hive now?

___ bees

Problem Solving • Applications  WRITE Math

Solve. Draw or write to show your work.

8. **THINK SMARTER** There are 38 ants on a rock. 10 move to the grass. 10 walk up a tree. How many ants are on the rock now?

_____ ants

9. **GO DEEPER** There are 27 birds at the park. 50 more birds come. Then 50 fly away. How many birds are at the park now?

_____ birds

10. **THINK SMARTER** Match the math sentences that count on and back by tens.

$$25 + 40 = ? \qquad 65 + 20 = ? \qquad 45 + 30 = ?$$

• • •

• • •

$$65 - 40 = ? \qquad 75 - 30 = ? \qquad 85 - 20 = ?$$

 TAKE HOME ACTIVITY • On slips of paper, write 36 + 40 and 76 − 40. Ask your child to explain how to use the hundred chart to count on and back by tens to find the sum and the difference.

Related Addition and Subtraction

Common Core COMMON CORE STANDARDS—1.NBT.C.4
Use place value understanding and properties of operations to add and subtract.

Use the hundred chart to add and subtract. Count on and back by tens.

1. 16 + 60 = _____

   76 − 60 = _____

2. 61 + 30 = _____

   91 − 30 = _____

3. 64 + 20 = _____

   84 − 20 = _____

1	2	3	4	5	6	7	8	9	10
11	12	13	14	15	16	17	18	19	20
21	22	23	24	25	26	27	28	29	30
31	32	33	34	35	36	37	38	39	40
41	42	43	44	45	46	47	48	49	50
51	52	53	54	55	56	57	58	59	60
61	62	63	64	65	66	67	68	69	70
71	72	73	74	75	76	77	78	79	80
81	82	83	84	85	86	87	88	89	90
91	92	93	94	95	96	97	98	99	100

Problem Solving Real World

Choose a way to solve. Draw or write to show your work.

4. There are 53 leaves in a tree. 20 leaves blow away. How many leaves are left in the tree?

   _____ leaves

5. **WRITE Math** Write a number sentence to subtract 3 tens from 93.

   _____

   _____

Lesson Check (1.NBT.C.4)

1. What is 78 − 20?
Write the number.

_____

1	2	3	4	5	6	7	8	9	10
11	12	13	14	15	16	17	18	19	20
21	22	23	24	25	26	27	28	29	30
31	32	33	34	35	36	37	38	39	40
41	42	43	44	45	46	47	48	49	50
51	52	53	54	55	56	57	58	59	60
61	62	63	64	65	66	67	68	69	70
71	72	73	74	75	76	77	78	79	80
81	82	83	84	85	86	87	88	89	90
91	92	93	94	95	96	97	98	99	100

2. What is 37 + 50?
Write the number.

_____

Spiral Review (1.OA.A.1, 1.OA.B.3)

3. Use the model.
What is the difference of 7 − 3?
Write the number.

$$7 - 3 = \underline{\quad}$$

4. What is the sum for 0 + 7?
Write the number.

_____

FOR MORE PRACTICE
GO TO THE
Personal Math Trainer

Name _____

Practice Addition and Subtraction

Essential Question What different ways can you use to add and subtract?

Common Core | Number and Operations in Base Ten—
1.NBT.C.4, 1.NBT.C.6 *Also 1.OA.C.6*
MATHEMATICAL PRACTICES
MP1, MP2, MP3, MP8

Listen and Draw Real World

Draw to show the problem.
Then solve.

Math Talk
MATHEMATICAL PRACTICES
Describe How did you solve the problem?

FOR THE TEACHER • Read the following problem. The class collects paper bags for an art project. Ron brings 7 more bags than Ben. Ben brings 35 bags. How many bags does Ron bring?

Model and Draw

What ways have you learned to add and subtract?

$5 + 9 =$ ___

> **THINK**
> 9 + 5 is the same as 10 + ___ .

$50 - 30 =$ ___

> **THINK**
> 5 tens − 3 tens.

$51 + 21 =$ ___

> **THINK**
> 5 tens + 2 tens.
> 1 one + 1 one.

Share and Show MATH BOARD

Add or subtract.

1. $30 + 60 =$ ___

2. $73 + 5 =$ ___

3. $10 - 4 =$ ___

4. $29 + 4 =$ ___

5. $9 + 9 =$ ___

6. $5 + 6 =$ ___

7. $25 + 54 =$ ___

8. $15 - 8 =$ ___

9. $40 + 10 =$ ___

10. $40 - 10 =$ ___

11. $14 - 7 =$ ___

12. $90 - 70 =$ ___

13. $86 + 12 =$ ___

14. $1 + 9 =$ ___

15. $6 + 7 =$ ___

16. $9 - 2 =$ ___

17. ✓ $8 + 31 =$ ___

18. ✓ $50 + 11 =$ ___

Name _____

On Your Own

Use Repeated Reasoning Add or subtract.

19. $\begin{array}{r} 12 \\ -\ 3 \\ \hline \end{array}$	20. $\begin{array}{r} 10 \\ +10 \\ \hline \end{array}$	21. $\begin{array}{r} 7 \\ +42 \\ \hline \end{array}$	22. $\begin{array}{r} 41 \\ +36 \\ \hline \end{array}$
23. $\begin{array}{r} 8 \\ +10 \\ \hline \end{array}$	24. $\begin{array}{r} 16 \\ +\ 7 \\ \hline \end{array}$	25. $\begin{array}{r} 6 \\ -6 \\ \hline \end{array}$	26. $\begin{array}{r} 3 \\ +8 \\ \hline \end{array}$
27. $\begin{array}{r} 64 \\ +\ 3 \\ \hline \end{array}$	28. $\begin{array}{r} 60 \\ -30 \\ \hline \end{array}$	29. $\begin{array}{r} 2 \\ +7 \\ \hline \end{array}$	30. $\begin{array}{r} 5 \\ -1 \\ \hline \end{array}$
31. $\begin{array}{r} 13 \\ -\ 5 \\ \hline \end{array}$	32. $\begin{array}{r} 52 \\ +40 \\ \hline \end{array}$	33. $\begin{array}{r} 3 \\ +2 \\ \hline \end{array}$	34. $\begin{array}{r} 30 \\ +50 \\ \hline \end{array}$

Solve. Write or draw to explain.

35. **THINK SMARTER** Lara collects 8 more stamps than Samson. Samson collects 39 stamps. How many stamps does Lara collect?

_____ stamps

Problem Solving • Applications

 WRITE Math

Solve. Write or draw to explain.

36. THINK SMARTER Jane drew some stars. Then she drew 9 more stars. Now there are 19 stars. How many stars did Jane draw first?

_____ stars

37. THINK SMARTER Adel drew 10 more stars than Charlie. Charlie drew 24 stars. How many stars did Adel draw?

_____ stars

38. GO DEEPER Write three ways to get a sum of 49.

__ ◯ __ = 49

__ ◯ __ = 49

__ ◯ __ = 49

39. THINK SMARTER Find the sum of 23 and 30. Use any way to add.

$$23 + 30 = \underline{\quad}$$

Explain how you solved the problem.

TAKE HOME ACTIVITY • Have your child explain how he or she solved Exercise 36.

Practice Addition and Subtraction

Common Core
**COMMON CORE STANDARDS—1.NBT.C.4,
1.NBT.C.6** *Use place value understanding
and properties of operations to add and subtract.*

Add or subtract.

1.   20
  + 20
	2. 90
	  − 30
		3. 52
		  + 4
			4. 62
			  + 21
				5. 39
				  − 10

6.   8
  + 2
	7. 47
	  + 34
		8. 4
		  − 0
			9. 49
			  − 6
				10. 64
				  + 30

Problem Solving

Solve. Write or draw to explain.

11. Andrew read 17 pages of his book
before dinner. He read 9 more pages
after dinner. How many pages did
he read?

_____ pages

12. **WRITE Math** Write two ways
you could use to find 5 + 8.

_____

_____

_____

Lesson Check (1.NBT.C.4, 1.NBT.C.6)

1. What is the sum of 20 + 18?
   Write the sum.

$$20 + 18 = \underline{\quad}$$

2. What is the difference of 90 − 50?
   Write the difference.

$$90 - 50 = \underline{\quad}$$

Spiral Review (1.OA.A.1, 1.OA.C.6)

3. Use the model. What number
   sentence does this model show?
   Write the number sentence that
   the model shows.

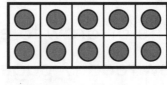

$$\underline{\quad} + \underline{\quad} = \underline{\quad}$$

4. Solve. Mo had some toys. He gave 6 away.
   Now he has 6 toys. How many toys did
   Mo start with?

$$\underline{\quad} \text{ toys}$$

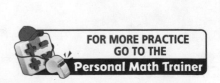

FOR MORE PRACTICE
GO TO THE
Personal Math Trainer

✔ Chapter 8 Review/Test

Personal Math Trainer
Online Assessment
and Intervention

I. Write each addition or subtraction problem
in the box below the answer.

| 7 + 2 | 3 + 3 | 15 − 9 | 8 + 6 | 14 − 5 |

6	9	14

---

2. Choose all the ways that name the model.

|| |||

○ **2** tens and **3** tens
○ **20 + 30**
○ **5**
○ **50**

---

3. Sasha has 70 stickers. She uses 40 of them.
How many stickers are left? Show your work.

_____ stickers

4. THINK SMARTER + Use the hundred chart to add.
Count on by ones or tens.

$37 + 5 =$ _____

Explain how you used the chart to find the sum.

1	2	3	4	5	6	7	8	9	10
11	12	13	14	15	16	17	18	19	20
21	22	23	24	25	26	27	28	29	30
31	32	33	34	35	36	37	38	39	40
41	42	43	44	45	46	47	48	49	50
51	52	53	54	55	56	57	58	59	60
61	62	63	64	65	66	67	68	69	70
71	72	73	74	75	76	77	78	79	80
81	82	83	84	85	86	87	88	89	90
91	92	93	94	95	96	97	98	99	100

5. Use the model. Draw to show how to add the tens.

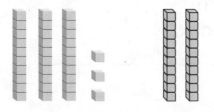

$33 + 20 =$ _____

Name _____

6. Use the model. Draw to show how to make a ten.

26 + 7 = ____

7. Write the addition sentence that the model shows. Solve.

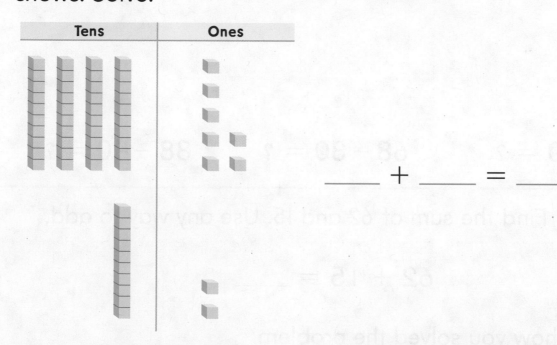

____ + ____ = ____

8. What is the difference?

15
− 7
_____

○ 7   ○ 8   ○ 10   ○ 12

9. What is the sum?

40
+ 50
_____

○ 10   ○ 70   ○ 80   ○ 90

10. Luis has 16 🍂 .

He has 38 🍂 .

How many leaves does Luis have? Circle the number that makes the sentence true.

Luis has $\begin{array}{|c|} 48 \\ 54 \\ 59 \end{array}$ leaves.

11. Match the math sentences that count on and back by tens.

$$38 + 30 = ?$$ $$48 + 40 = ?$$ $$38 + 20 = ?$$

•   •   •

•   •   •

$$58 - 20 = ?$$ $$68 - 30 = ?$$ $$88 - 40 = ?$$

12. GO DEEPER   Find the sum of 62 and 15. Use any way to add.

$$62 + 15 = \underline{\qquad}$$

Explain how you solved the problem.